THE LAST STRAW

Kids vs. Plastics

Written by SUSAN HOOD

Illustrated by CHRISTIANE ENGEL

HARPER

An Imprint of HarperCollinsPublishers

For Emily, Allison, Sophie, Molly, and all the kids
who make our world a better place.
—S.H.

To all those awesome ecowarriors out there!
—C.E.

Contents

Introduction

Hi! My name's Milo.

When I was nine years old, I noticed something: whenever I ordered a drink, it came with a plastic straw, whether I wanted one or not.

This seemed like a huge waste, because I don't usually need a straw. It made me wonder how many straws are used each year. Too many! Lined up end to end, the straws we use in America each year would circle the Earth twice!

That's when I decided I can do something about this. Kids can do something about this. Anyone who can order a drink can order one without a straw if they don't need one. More and more people are waking up to the absurdity of using a new straw for every drink, then throwing it away fifteen minutes later, hoping or imagining that it will vanish into thin air.

Every bit of plastic waste we throw away today will be here on Earth, somewhere on Earth, in a landfill, in a field, in the ocean or a stream, long after my great-grandchildren—and yours—are born. They'll be looking at the mountains of plastic straws we throw away today and wondering why there are so many, and why we didn't do things differently, and how we could have left such a mess for them to clean up.

That's just not a legacy I want to leave behind. So I started the Be Straw Free project, which gave me a voice for change—on a local and global scale.

This project was something that I wanted to share with other kids, and with the help of some generous supporters, I went on an international speaking tour to talk to kids about finding and getting involved with projects that interest them.

I went on this tour because, well, here's the thing . . . this planet's not a place that kids will inherit at some point far off in the distant future. We live here right now; we can and, in fact, do effect change already.

What I found, everywhere I went, is that kids around the world are coming up with creative ways to live on this planet and to take care of this planet and each other. And that's what this book by Susan Hood is all about.

The thing is, plastic straws are just the beginning. When I first started this project, I thought that adults wouldn't listen to what a kid has to say. But I found that being a kid can actually be an advantage. People are willing to listen to your ideas. And when you've gotten people to make a small change in their life, like using fewer plastic straws, or bags, or bottles, it makes them think about all the different ways they can make a change in their communities.

You can change your community, your town, your state. You can change the world! So c'mon, let's do this!

—Milo Cress
Founder, BeStrawFree.org

Fantastic Plastic

Is plastic fantastic?
No doubt about it!
Where in the world
would we be without it?

Where in the world
would hospitals be
if medical tools
weren't germ-free?

How in the world
would we safely play sports
without proper gear
on our fields and our courts?

What in the world
would we want to eat
if food wasn't fresh
in the store down the street?

Where in the world
would technology be?
Think smartphones, computers!
See? Plastic is key!

Is plastic a blessing?
Or is it a curse?
It makes our lives better.
BUT could they get worse?

Imagine medicine without disposable surgical gloves,
IV bags, syringes, pacemakers, stents; sports without
helmets, goggles, mouth guards, life vests; grocery
stores without protective food wrapping. How do we
use—and reuse—the plastic we need, refuse the plastic
we don't, and avoid abusing the Earth?

7

P Is for Peek-a-Boo Plastic

Take a peek where plastic can hide!
Would you believe there is plastic inside . . .

Athletic apparel

Eyeglasses

Fishing line
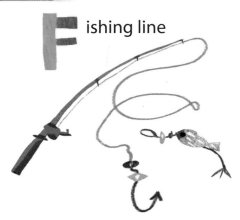

Glitter

Kites

Lint

Money

Quilting thread

Ribbons

Sneakers

Wrapping paper

X-rays
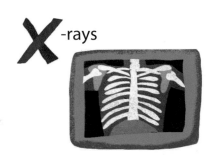

Yogurt containers

8

B 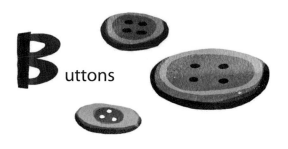uttons

C hewing gum

D isposable diapers

H elmets

I ce cream cups

J uice boxes

N onstick pans

O ven gloves

P roduce stickers

T ires

U mbrellas

V elcro

Z ippers

In 2017, scientists estimated that we've manufactured 8.3 billion metric tons of plastic to date—the weight of 25,000 Empire State Buildings or 1 billion elephants!

"Only nine percent of plastic is recycled."
—Roland Geyer, industrial ecologist, UC Santa Barbara, California, United States

A Sea Change

Listen to the seagulls cry,
watching whales
who used to thrive
in seas of cobalt blue.
Those mighty mammals ruled the waves—
a most majestic crew!

Listen to the seagulls cry,
watching whales
who breach and dive
in seas of plastic stew.
Whales eat their fill of bags and cups
and other human spew.

Listen to the seagulls cry,
watching whales
who can't survive
sink slowly out of view . . .

O wisest of the mammals, please!
Sea change is up to you.

In 2019, a young whale washed ashore in the Philippines with eighty-eight pounds of plastic bags, nylon ropes, and rice sacks in its stomach. The plastic clogged its digestive system, and feeling full, the whale didn't eat. With no way to digest or expel the plastic, it died of starvation and dehydration.

"The [whales] that land on the beach are . . . just the tip of the iceberg. They're just the ones we see."
—Lars Bejder, director of the Marine Mammal Research Program, University of Hawaiʻi, United States

Plastic for Dinner?

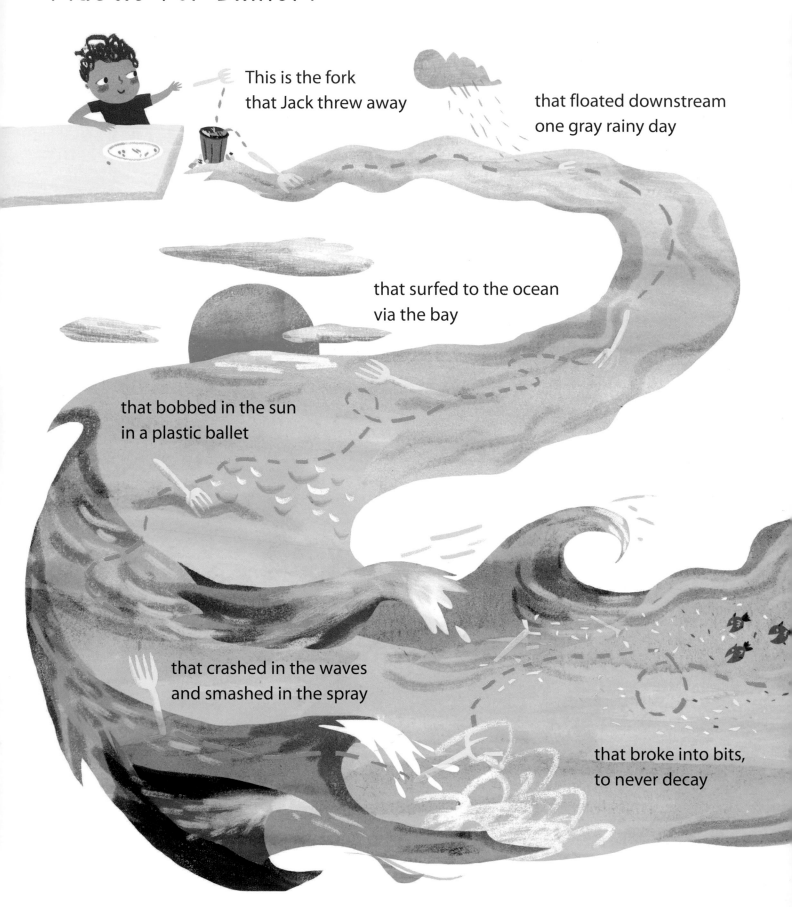

This is the fork
that Jack threw away

that floated downstream
one gray rainy day

that surfed to the ocean
via the bay

that bobbed in the sun
in a plastic ballet

that crashed in the waves
and smashed in the spray

that broke into bits,
to never decay

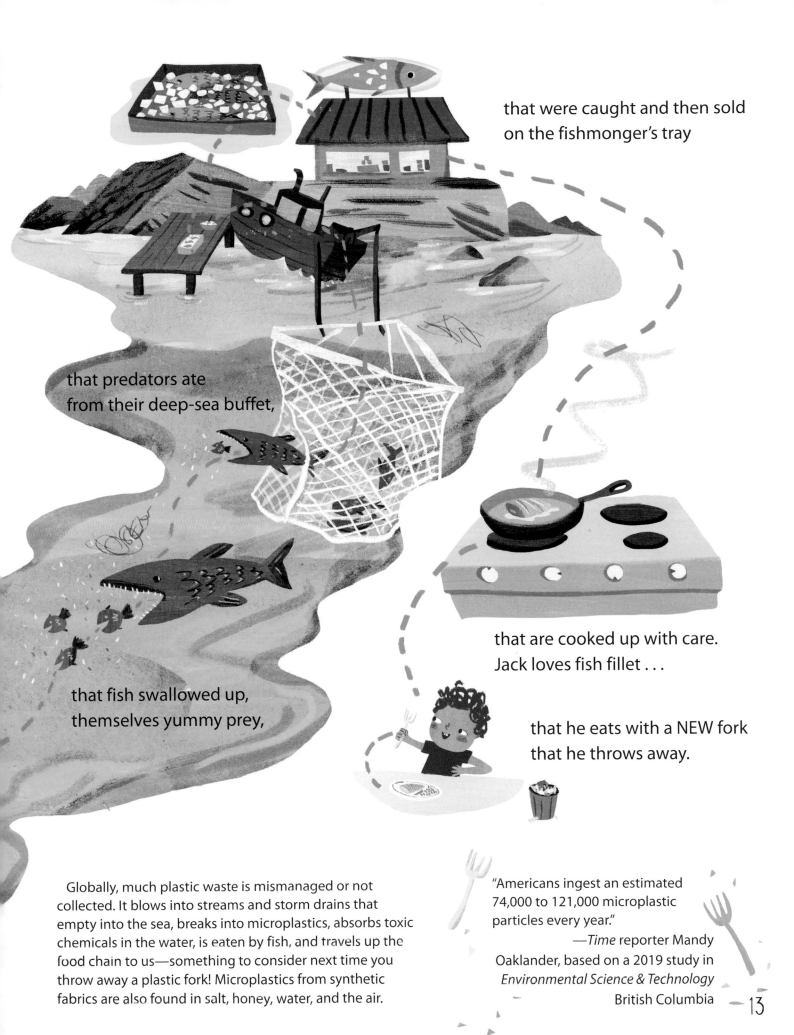

that were caught and then sold
on the fishmonger's tray

that predators ate
from their deep-sea buffet,

that are cooked up with care.
Jack loves fish fillet . . .

that fish swallowed up,
themselves yummy prey,

that he eats with a NEW fork
that he throws away.

Globally, much plastic waste is mismanaged or not collected. It blows into streams and storm drains that empty into the sea, breaks into microplastics, absorbs toxic chemicals in the water, is eaten by fish, and travels up the food chain to us—something to consider next time you throw away a plastic fork! Microplastics from synthetic fabrics are also found in salt, honey, water, and the air.

"Americans ingest an estimated 74,000 to 121,000 microplastic particles every year."
—*Time* reporter Mandy Oaklander, based on a 2019 study in *Environmental Science & Technology*

British Columbia

13

The Great Pacific Garbage Patch

Round and round and round it flows, and every year it grows

We need an eco-revolution! •

We need a plan, a new solution. We need to toxic goop.

NORTH PACIFIC GYRE

Pacific Ocean

WESTERN GARBAGE PATCH

fishing nets, stirrers, straws, and cigarettes. Broken bits, confetti soup, turn the seas to toxic goop.

Pushed by the wind and circling currents called gyres, submerged trash (which some estimate to be more than twice the size of Texas) swirls clockwise around the North Pacific Ocean. More like a trash soup than an island, it concentrates in two dense spirals. The microplastics' bright colors attract fish, birds, and other sea life, causing them to choke and die.

Fionn Ferreira, an eighteen-year-old from Ireland, won the 2019 Google Science Fair for a promising chemical solution that removed most microplastics from his water test sample. But only time will tell if his idea will work on a larger scale.

and grows. Pounds and pounds of trashed debris, a monster vortex swirls the sea. Like the other ocean gyres, the great Pacific reels with tires, bottles, toys, old

EASTERN GARBAGE PATCH

"I was confronted, as far as the eye could see, with the sight of plastic."
—Sailboat captain Charles Moore upon discovering the Great Garbage Patch in 1997

Ban the Bag

Say no, no thank you, not today.
No, not next week or any day!

Plastic bags should all be banned.
Tell your stores to take a stand.

They're handy, sure, but what's the cost
if plastic bags mean lives are lost?

Bags float on air and snag the trees,
they swirl on currents round the seas.

Sea turtles spy a favorite dish—
mistaking bags for jellyfish!

They eat them up, they get their fill,
and starve to death. Those bags can kill.

Plastic never disappears
but breaks apart for years and years.

Don't let plastic drag us down.
Take your own tote bag to town.

Cloth and canvas can't be beat.
They make plastic obsolete.

Tell your stores to take a stand.
Plastic bags should all be banned!

Americans use about 100 billion plastic shopping bags every year. If you counted these bags one bag per second, it would take you (and your descendants!) nearly 3,200 years! Plastic doesn't decompose; it just breaks apart into smaller and smaller bits over the course of 450 to 1,000 years. So nearly every piece of plastic that was ever created still exists.

Mr. Trash Wheel

There once was a googly-eyed guy
who gobbled the trash floating by.
The Baltimore river
would always deliver.
It seemed like an endless supply!

One birthday, kids made him a cake
of plastic and trash, no mistake!
They brought it to town.
He swallowed it down.
Did it give him a big bellyache?

No!

Citywide, Mr. Trash is adored
for gulping this gross smorgasbord.
He keeps trash away
from fish in the bay.
He deserves a clean-ocean award!

Rivers are pouring "between 1.15 and 2.41 million metric
tons per year of plastic waste into the world's ocean."
—Oceanographer Laurent Lebreton,
Waikato, New Zealand

(That's like dumping 85,000–180,000
school buses into the ocean!)

Since his 2014 installation, Mr. Trash Wheel has gobbled up more than 1,000 tons of trash floating down the Jones Falls River, preventing it from flowing into the Baltimore harbor and the ocean.

Kids from Waverly Elementary School and Commodore John Rodgers Elementary School celebrated Mr. Trash Wheel's birthday by feeding him a trash cake. Statistics on items collected by Mr. Trash Wheel helped a student-led fight against Styrofoam that resulted in one of the first statewide bans.

The Road Back

Sokanha Ly and Bunhourng Tan are paving the way to reuse the discarded plastic overwhelming Cambodia, which has no recycling facilities. They propose using plastic to build much-needed roads. The process would mix plastic scraps with bitumen to make plastic asphalt concrete, which is cheaper and stronger than traditional asphalt. (Note: Some worry that plastic roads may introduce more microplastics into the soil.)

From Bottles to Buddies

Some bullies can often be cruel.
Young Sammie knew that wasn't cool.
She pulled out the stops
to melt bottle tops
for a buddy bench for her school.

Now kids who are lonely or shy
can signal to those passing by:
If you want to come play,
please sit here today.
A friendship begins saying "Hi!"

"If someone is lonely they can go
sit on the bench and others know to
go up and ask them to play."

—Sammie Vance (age 9),
Indiana, United States

Sammie Vance collected more than 1,600 pounds of bottle caps
from all fifty states as well as from Africa, Israel, Germany, Mexico, and
the Netherlands. (Her pile would weigh more than eight 200-pound
humans!) She delivered those caps to a company that melted them
down to create three buddy benches for her school and helped
neighboring schools that wanted to follow Sammie's lead.

For the Love of Frogs

This is the boy
who sold some toys,
all for the love of frogs.

He raised some cash
and picked up trash,
all for the love of frogs.

He spread the word.
His city heard!
All for the love of frogs.

They cleaned the pond
and tadpoles spawned.
All for the love of frogs.

That was the start—
he gave his heart,
all for the love of frogs.

Dismayed to hear that plastic pollution was killing his favorite animal, Justin Sather raised more than $1,000 to clean up frog habitats with the help of Save the Frogs! After hearing about Sammie Vance (see opposite page), Justin collected 200 pounds of plastic bottle caps to recycle into a buddy bench for his school.

"Frogs are telling us the world needs our help."
—Justin Sather (age 8), California, United States

23

Ode to the Jellyfish

Crystal bells, pulse and swish,
all hail the stunning jellyfish!
Boneless, spineless,
brainless, bloodless,
they've bloomed across seas
endlessly, for centuries.
Elegant swimmers,
these see-through skimmers
may be the solution
to plastic pollution!

The key? Jellyfish snot,
believe it or not!
Icky, slicky,
tricky, sticky!
Scientists' new focus?
Mucus hocus-pocus
filtering the ocean—
a spellbinding potion!
What's the secret in the snot
of this ancient Argonaut?

Dive in, Science, swim and swish.
All hail the stunning jellyfish!

Dr. Dror Angel, marine ecologist at the University of Haifa in Israel, along with his international colleagues at GoJelly, is studying the use of jellyfish mucus to capture microplastics and reduce their flow to the ocean. Professor Yael Kali's fourth and fifth graders at the Rambam School in Nahariya, Israel, adopted Dr. Angel's jellyfish reporting website, www.meduzot.co.il, and are working to be Citizen Science Ambassadors, spreading the word about the amazing jellies!

What Can a Bottle Be?

Once I stood tall,
full to the brim,
flanked by my friends,
row upon row.
Where would we go?
What would we be?

Then, suddenly,
I was turned
upside down,
left empty, alone,
all on my own.
Where would I go?
What would I be?

Week after week,
blown far from home,
sniffed by stray dogs,
kicked down the paths,
forgotten, trashed.
Where would I go?
Is this all I would be?

One sunny day,
an amigo
reached out to me.
"Perfecto," he said.
Does he mean me?
Where would we go?
What would I be?

Now I stand tall,
full to the brim,
flanked by my friends,
some old, some new,
row upon row.
Now we all know
what we can be.
Something so cool—
a school!

"By building bottle schools in Guatemala, entire communities come together to clean up their environment . . . and kids get to say, 'I helped build my own school!'"

—Adam Flores, Hug It Forward, San Martín Jilotepeque, Chimaltenango, Guatemala

Children in Guatemala collect discarded plastic bottles, stuff them with other litter, and use them as building blocks called ecobricks to create walls for badly needed new schools. Ecobricks are cheaper than cement blocks and act as good insulation.

27

The Munching, Crunching Caterpillars

The munching, crunching caterpillars
lived in a beehive
feeding on beeswax—

munch, munch, munch.

The very tall beekeeper
plunked those pesky caterpillars
into a plastic bag—

scrunch, scrunch, scrunch.

The munching, crunching caterpillars
lived in the bag,
feeding on plastic—

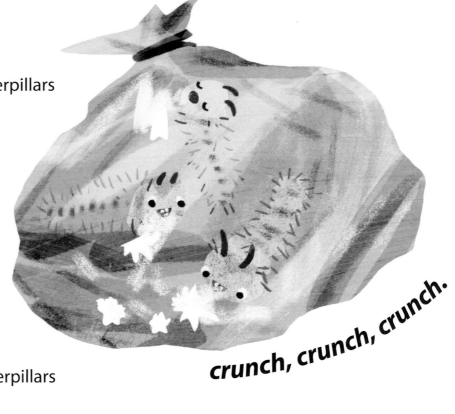

crunch, crunch, crunch.

The munching, crunching caterpillars
did something fantastic.
They digested plastic,
making it disappear!
The munching, crunching caterpillars
gave scientists a

hunch, hunch, hunch!

Sometimes the smallest among us
can help solve our biggest problems.

"The hunt for
organisms that can
degrade plastics is on."
—Jennifer DeBruyn,
University of Tennessee,
United States

Federica Bertocchini, a scientist and beekeeper in Spain, pulled wax moth caterpillars out of her beehive and put them in a plastic bag. Forty minutes later the bag was full of holes. Some chemical inside the caterpillars' digestive system was breaking down the plastic. Much work is still to be done, but the caterpillars may help us inch along to a solution for plastic pollution.

29

Be Straw Free

We owe our thanks to Milo Cress,
a boy just nine years old.
He noticed straws used in excess.
We owe our thanks to Milo Cress.
His big idea's a huge success—
"Be Straw Free" is taking hold!
We owe our thanks to Milo Cress,
a boy just nine years old.

In 2011, fourth grader Milo Cress helped kick-start a movement when he asked a local Vermont café to offer straws only if customers asked for them. Milo doesn't advocate a ban on straws, because many people with disabilities require straws to drink. But drastically cutting straw usage is cheaper for restaurants and reduces waste. Thanks to Milo's idea, major companies are going straw free.

A Shining Light

Smart girl
uses plastic
plus sunlight's pure power
to heat hot water for showers.
Brilliant!

"I want . . . to install my solar heater in homes of indigenous communities in Chiapas so that people will no longer get sick . . . [from bathing] with cold water."
—Xóchitl Guadalupe Cruz López (age 8),
San Cristóbal, Chiapa, Mexico

Young Xóchitl invented a rooftop solar heater with recycled plastic bottles, hose, cable ties, nylon, and glass to provide hot water for bathing and won a science prize from the Mexican National Autonomous University—an award that is usually given to adults.

Stand Up, Speak Up

Fed up?

Get up. Rev up. Join up!

Clean up. Dream up.

Goof up? Fix up!

Stand up.

Shake

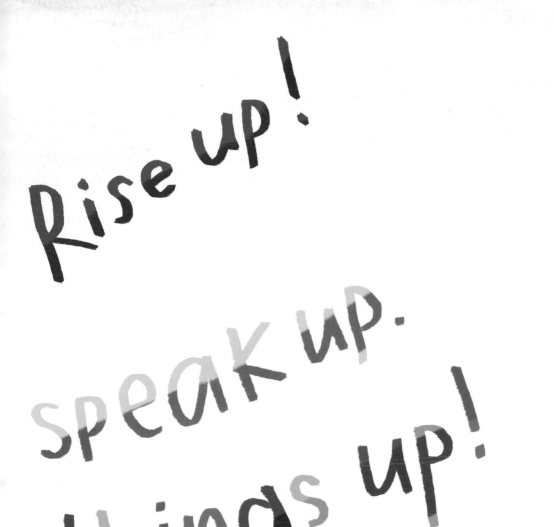

Rise up!
speak up.
things up!

Just don't give up!

"Spread the word. Post on social media, tell the brands you buy what you want them to do differently, and talk to your family and friends about why they should refuse plastic."
—Amy Meek (age 13),
Nottingham, United Kingdom

Sisters named Amy and Ella Meek spoke up to the then-British prime minister, Theresa May, saying that the UK's plan to cut all avoidable plastic by 2042 wasn't "drastic or urgent enough." They won a Points of Light Award, and their campaign was discussed in the House of Commons, the Scottish Parliament, and the Welsh Assembly!

Join the Crew

We may think
the problem's too big
and we are too small.
Just kids after all.
What can we do . . . ?

I can pick up a bottle, a flip-flop, a straw.
And you can too. And you. And You. And YOU!

Many hands, many faces,
many lands, many races
can scrap the old ways
and piece together a craft—
an ark, a lifeboat,
a patchwork of many colors—
to sail the seven seas,
to tell the world
this is our home
and it's worth changing course
to protect it.

There's rough water ahead—
riptides, dire straits—
but we can come about,
trim our sails,
and follow our North Star.
Our lives depend on it.

Come, step aboard.
Plot your course.
Take the helm—
it's your watch.

THIS VOYAGE BEGINS WITH YOU.

In 2016, Ben Morison, Ali Skanda, and Dipesh Pabari built a traditional Swahili dhow from more than 7 metric tons of waste plastic (heavier than an African elephant) found on a 10-kilometer (6-mile) stretch of the Kenyan coast. More than 3,000 students helped collect the plastic, which was melted down into plastic beams and planks. Clad in 30,000 recycled flip-flops, the *Flipflopi* sailed along the coast of East Africa in 2018, calling attention to plastic pollution and organizing coastal cleanups.

"Plant that one tree, pick up that one plastic bottle, the little things matter. . . . Everyone is able to make an impact and bring about change."
—Charlotte Wanja (age 17), speaking at the 2019 Fourth United Nations Environment Assembly, Nairobi, Kenya

Author's Note

As a sailor who has loved the ocean all my life, I was shocked by news of a whale in the Philippines starving to death after ingesting eighty-eight pounds of plastic. For me, it was a long-overdue wake-up call. Soon I started seeing plastic everywhere, as indeed it is: in water bottles, takeout containers, grocery bags, toothbrushes, and much more. Did you know there's plastic in gum? In glitter? In lint? In the air?

Don't get me wrong: Plastic *is* a miraculous invention. Think of plastic airbags, stents, MRIs, cell phones, computers, sports safety equipment. It's convenient, lightweight, inexpensive, sometimes lifesaving . . . but it's piling up everywhere. I'm as guilty as the next person of using too much plastic.

Until now, I had reassured myself about my own plastic consumption because my family recycles. But a groundbreaking study by scientists at UC Santa Barbara says 91 percent of plastic isn't recycled. *What?* China used to handle nearly half of the world's recycling, including waste from the United States, but no more. Other countries have no recycling facilities, or the process is complex and costly. Different plastics cannot be recycled together; one mismatch can contaminate the whole batch. Some items, like plastic bags, can clog the machinery at recycling facilities, causing shutdowns, which take time—and money—to fix. Plastics can rarely be recycled into the same product; the process changes the plastic's chemistry, so it's *downcycled* into different things like carpet or doormats. And at its core, new plastic is inexpensive. So it's cheaper for companies to create brand-new plastics to fill the demand for disposable water bottles, bags, straws, and so on. At such low rates, recycling isn't contributing to stemming the flow of new plastics flooding the Earth, with dire consequences for us all. The facts are alarming:

- Researchers estimate we've created more than 8.3 billion metric tons of plastic since the 1950s. According to the University of Georgia, that's the weight of 80 million blue whales!

- It takes decades or even centuries for plastic to break down, sometimes as many as 450–1,000 years. Scientists think it never really decomposes, but just fragments into smaller and smaller bits.
- Where does it go? As of 2015, 9 percent was recycled, 12 percent was incinerated (releasing toxic gases), and 79 percent went into landfills or the environment. More than 8 million metric tons of plastic end up in the oceans *every year*. That's 8.8 million US tons (more than 1½ times the weight of the Great Pyramids of Khufu in Egypt or more than 44,444 blue whales). This plastic causes the deaths of a million seabirds and more than 100,000 marine mammals.
- As it breaks down, plastic can leach chemicals into the land and water. Animals eat them, then these chemicals travel up the food chain to us.
- In the sunlight, plastics on land release greenhouse gases like methane and ethylene that cause climate change.
- Without a change in our ways, there's no end in sight. Plastic production is predicted to triple by 2050.

However, there is good news. While researching this book, I was thrilled and inspired by the children and teenagers who are tackling this problem head-on. By 2050, when these kids are adults, there could be more plastic in the ocean (by weight) than fish. For them, the fight is personal.

What can *I* do? A good place to start is to fight single-use plastics—the use-it-once-and-throw-it-away plastic straws, water bottles, disposable shopping bags, coffee cups, and utensils that make up 40% of all plastic waste. If I can rethink and replace these things with reusable items (metal water bottles, cloth bags, bamboo utensils); if I can refuse things I don't need; if I can help remove plastic litter; if I can encourage innovative research; if I can help report the problem, and if you can too, the world will be a better place for our kids . . . and theirs.

—Susan Hood

Use This, ~~Not Plastic~~

Single-use plastics like many of the items below are the biggest culprits in plastic pollution. We use them for minutes, but they can last for centuries. There are eco-friendly alternatives!

Glass jar
~~Plastic jar~~

Metal, glass, or bamboo straw
~~Plastic straw~~

Bamboo toothbrush
~~Plastic toothbrush~~

Metal water bottle
~~Plastic water bottle~~

Ice cream cone
~~Ice cream cup~~

Reusable lunch bag
~~Plastic lunch bag~~

Wooden clothespin
~~Plastic chip clip~~

Wooden comb
~~Plastic comb~~

Beeswax cotton wrap
~~Plastic wrap~~

String bag, cloth bag, or basket
~~Plastic grocery bag~~

Bamboo knives, forks, spoons
~~Plastic knives, forks, spoons~~

Cotton, wool, hemp, and silk clothes
~~Synthetic clothes~~

Top Ten Ocean Polluters

Here are the top ten most commonly found items on beaches and in waterways, according to Ocean Conservancy's 2018 International Coastal Cleanup, conducted in more than a hundred countries.

We throw out so much plastic, it's "like dumping a . . . garbage truck full of plastic into the ocean every minute of every day for an entire year!"
—Ocean Conservancy

#1
Cigarette butts
(plastic filters)

#2
Food wrappers

#3
Plastic straws and
stirrers

#4
Plastic forks,
knives, and spoons

#5
Plastic beverage
bottles

#6
Plastic bottle caps

#7
Plastic grocery
bags

#8
Other plastic bags

#9
Plastic lids

#10
Plastic cups and
plates

Sources and More

At the time of publication, all website links were current. For clickable links, visit www.harpercollins.com/thelaststraw.

Fantastic Plastic

Plastics have fueled health and safety advances and made vehicles lighter and more fuel efficient, reducing greenhouse gas emissions. They insulate buildings, making them more energy efficient. And they help build solar panels, wind turbines, and solar water heaters.

The good news about plastics

www.plasticsmakeitpossible.com
www.chemicalsafetyfacts.org/plastics

P Is for Peek-a-Boo Plastic

Write a list of the plastics you use in twenty-four hours. Or collect the plastic you use in a week. You'll be shocked at the size of the pile. Or try to go one day without using plastic. Bet you can't!

Quote: www.advances.sciencemag.org/content/3/7/e1700782.full

Facts about our plastic consumption: www.sciencedaily.com/releases/2017/07/170719140939.htm

A Sea Change

About half of the plastic in the ocean enters from Asia, especially from China, Indonesia, and the Philippines. To be fair, much of that plastic waste was shipped to Asia from the United States, Canada, Australia, and other countries to be recycled. Now Asian countries like China and Malaysia are sending it back and not accepting new shipments.

Quote: www.vox.com/2019/5/24/18635543/plastic-bags-whale-stomach-beached
Reports about whales ingesting plastic:
www.nytimes.com/2019/03/18/world/asia/whale-plastics-philippines.html?action=click&module=Latest&pgtype=Homepage&fbclid=IwAR3ejeCg-GVhLRrLiHTOP1lygc60ez9HYmg1eOeUhmzyOmS3PzIisUHd7b4
Plastic in Asia: www.latimes.com/world/la-fg-asia-plastic-waste-20190617-story.html
www.oceanconservancy.org/wp-content/uploads/2017/04/full-report-stemming-the.pdf

Plastic for Dinner?

Microplastics are less than 5 millimeters long (1/5 of an inch), about the size of a sesame seed or a black ant. They can be eaten by animals and travel up the food chain to humans. We also inhale microplastics floating in the air. Those who drink bottled water can get 86,000 more microplastics per year than those who drink from the tap. The health consequences have yet to be determined.

Quotes: www.time.com/5601359/microplastics-in-food-air
https://pubs.acs.org/doi/10.1021/acs.est.9b01517
Microplastics and human health:
www.ncbi.nlm.nih.gov/pmc/articles/PMC6132564

The Great Pacific Garbage Patch

Garbage patches are developing in the Indian Ocean, the North and South Atlantic, and the North and South Pacific, concentrated by rotating ocean currents called gyres. Scientists estimate there are about 5.2 trillion plastic fragments in the five gyres. To clean it up, you and the other 7.7 billion people on Earth would have to pick up about 675 pieces each. Tiny Henderson Island in the South Pacific is uninhabited and far from humans, but thanks to the gyres, it's one of the most polluted places on Earth.

Quote: www.onlinegreatpacificgarbagepatch.weebly.com/discovery.html
Sailing the garbage patch: https://blog.nationalgeographic.org/2017/07/28/charlesmoore-is-now-a-two-time-garbage-patch-discoverer-and-i can-tell-you-what-a-garbage-patch-looks-like
Fionn Ferreira's project: www.thejournal.ie/irish-student-science-award-microplastics-4745270-Jul2019/?fbclid=IwAR0XFZHtTOuHURq01nZBmFgvlQ8uld1gX-74mW-oMuMqQr-eBBnfYthWl4M

Ban the Bag

Single-use plastic bags are created using oil or natural gas that took millions of years to form, and we throw the bags away after about twelve minutes. (The energy needed to create twelve plastic bags could fuel a car for one mile!) These bags choke sea life, cattle, sheep, and camels, and they are rarely recyclable because they clog the machinery. Countries, states, and cities are working to ban them or charge fees to discourage their use. Note: Paper grocery bags require even more energy to produce, so the best option is a reusable bag that you can wash in the laundry.

Quote: *Plastic: A Toxic Love Story* by Susan Freinkel, p. 277
Key facts: www.biologicaldiversity.org/programs/population_and_sustainability/sustainability/plastic_bag_facts.html
Paper or Plastic? www.washingtonpost.com/wp-dyn/content/graphic/2007/10/03/GR2007100301385.html?referrer=emaillink

Mr. Trash Wheel

Mr. Trash Wheel, installed in 2014, is now one of three solar- and water-powered contraptions helping to keep Baltimore's waters trash free. Created by John Kellett, the wheel has generated interest around the world, including in India, Hawai'i, and Bali. To date, Mr. Trash has collected more than 11 million cigarettes, 1 million foam containers, 880,646 plastic bottles, a guitar, and a python! He even has his own Twitter account.

Quote: https://theoceancleanup.com/updates/quantifying-global-plastic-inputs-from-rivers-into-oceans
How Mr. Trash Wheel works: www.mrtrashwheel.com

The Road Back

The first plastic road was built in New South Wales, Australia. India has already paved 30,000 kilometers of road with plastic (that's more than 18,600 miles, or 7 times the distance between New York and Los Angeles as the crow flies). Other countries are taking notice. Ly and Tan's start-up company, Eco-Plastic, would not only clean up the tons of plastic littering Cambodia but also transform the potholed roads causing accidents every year.

Quote: www.channelnewsasia.com/news/asia/cambodian-women-on-green-mission-build-roads-with-plastic-waste-10396120
How Ly and Tan set up Eco-Plastic: www.phnompenhpost.com/business/cambodian-duos-eco-plastic-takes-second-bhutan-startup-challenge

From Bottles to Buddies

Sammie's buddy benches have attracted attention from Reese Witherspoon, *Martha Stewart Living*, the *Today* show, and more.

Quote: www.sammiesbuddybenchproject.com/sammiesstory
***Today* show video:** www.today.com/video/everyone-has-a-story-honors-woman-who-helped-put-buddy-benches-in-school-1241005123909

For the Love of Frogs

Justin Sather has presented his ideas to Dr. Jane Goodall at her Roots and Shoots program.

Quote: www.fortheloveoffrogs.com
Jane Goodall's Roots and Shoots recycling programs: www.rootsandshoots.org/projects/search?f%5B0%5D=im_field_rnsp_focus%3A7

Ode to the Jellyfish

Jellyfish can be invasive species, with tens of millions of them swarming Israel's coast in one week in 2019.

(That number equals more than all the people in New York City, Los Angeles, Chicago, Houston, Phoenix, and Philadelphia combined!)

Quote: www.israelhayom.com/2019/07/15/marine-biology-experts-unpack-israels-jellyfish-problem
Dr. Angel's research: www.nocamels.com/2018/08/jellyfish-plastic-waste-israeli-scientists
http://gojelly.eu

What Can a Bottle Be?

Ecobricks are the brainchild of environmentalist Susanna Heisse. Schools built from bottles clean up villages and spread self-reliance and pride as the whole community gets involved. The idea has now spread to the Philippines and South Africa.

Quote: Author interview with Adam Flores, Hug It Forward, www.hugitforward.org
All about ecobricks: www.theguardian.com/lifeandstyle/2014/may/29/ecobricks-and-education-how-plastic-bottle-rubbish-is-helping-build-schools and www.ecobricks.org

The Munching, Crunching Caterpillars

Other scientists from the United States, Japan, and China are working with various fungi and bacteria that can speed plastic's decomposition, given the right conditions.

Quote: www.theatlantic.com/science/archive/2017/04/the-very-hungry-plastic-eating-caterpillar/524097
Caterpillars and plastic: www.nature.com/articles/d41586-017-00593-y

Be Straw Free

Americans use an estimated 170–390 million straws *per day*! (Lined up end to end, 390 million straws would circle the Earth twice.) Neither plastic nor paper straws are accepted at most recycling facilities, because they drop through the sorters and can contaminate other plastics being recycled. For the disabled or others who may need straws, reusable ones are available made of stainless steel, glass, reusable plastic, or bamboo.

Quote: www.nps.gov/articles/straw-free.htm
CNN interview with Milo video:
www.youtube.com/watch?v=VtAjlU4-ffl
Straw FAQs: www.strawlessocean.org/faq

A Shining Light

Xóchitl's solar water heater can heat more than 10 liters (2.6 gallons) to 35-45°C (95-113°F), even in cold weather.
Quote: https://translate.google.com/translate?hl=en&sl=es&u=https://es-mb.theepochtimes.com/nina-mexicana-inventa-calentador-agua-ganando-prestigioso-premio-ciencias_514654.html&prev=search
Science Awards: www.remezcla.com/culture/8-year-old-girl-who-built-solar-water-heater-is-first-to-win-this-mexican-science-award

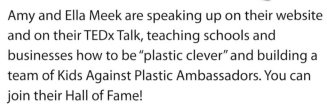

Stand Up, Speak Up

Amy and Ella Meek are speaking up on their website and on their TEDx Talk, teaching schools and businesses how to be "plastic clever" and building a team of Kids Against Plastic Ambassadors. You can join their Hall of Fame!

Quote: www.huffingtonpost.co.uk/maryann-ochota/kids-vs-plastic_b_15160036.html
Website: www.kidsagainstplastic.co.uk

Join the Crew

The *FlipFlopi* is a traditional Swahili dhow made of 100% recycled plastic, built to show how single-use plastics could have a second life and a positive impact.

Quotes: www.allafrica.com/stories/201903130520.html
All about the *Flipflopi*: www.theflipflopi.com

HEAVY FACT:
Internationally, most scientists report facts in metric tons (each about 2204 pounds, unlike the US ton, which is 2000 pounds).

Poetry Notes

The poems in this book use a variety of different poetry techniques and formats. Here's more information on each.

"Fantastic Plastic" is a *question poem* in which a series of questions builds to dramatic effect.

"P Is for Peek-a-Boo Plastic" is an *ABC poem* that uses words with initial letters from A to Z. A similar type of poem is an *acrostic*, in which the first letter of each line spells out the subject of the poem when read from top to bottom.

"A Sea Change" is an *elegy*, a poetry form used to praise or mourn the dead, so it often has a sad or somber tone. Elegies follow no set form.

"Plastic for Dinner?" is an abbreviated version of a *cumulative poem* in the spirit of "This Is the House That Jack Built." Cumulative poems use repetition and usually come full circle.

"The Great Pacific Garbage Patch" is a *concrete* or *shape poem*, where words are arranged to form a picture of the subject of the poem. It can rhyme or not.

"Ban the Bag" is a poem composed of *couplets*, stanzas made of two lines that rhyme.

"Mr. Trash Wheel" is a *limerick*; each stanza has five lines with a rhyme scheme of aabba. Popularized by Edward Lear, limericks predominantly use anapests, one form of poetic meter. An anapest has three syllables, with one strong beat at the end: soft/soft/STRONG. It can be found in a string of words such as "in the bay" or in a single word such as dis/ap/PEAR. There are three strong beats in lines one, two, and five, and two strong beats in lines three and four.

Mr. Trash Wheel gobbles trash floating down the Jones Falls River, protecting both the Baltimore, MD, harbor and the ocean beyond.
Photo credit: Waterfront Partnership of Baltimore

"The Road Back" is a *found poem*, which can take existing text found in signs, newspaper articles, graffiti, letters—any text—and use selected words to create a poem. The poet can add or delete text, change the lines or spacing, or leave the words unchanged. It's like a word collage.

"From Bottles to Buddies" is a *limerick*, as is "Mr. Trash Wheel."

"For the Love of Frogs" is a poem written in *tercets*, three-line stanzas that can rhyme or not. This one has a *refrain* (a repeating line) in the third line.

"Ode to the Jellyfish" is an *ode*, which means it celebrates a person, animal, or object. It often has no formal structure and may or may not rhyme.

"What Can a Bottle Be?" is a *persona poem*, which is written from the point of view of the poem's subject.

Once in place, bottles are covered with a layer of cement and the new walls are painted to create colorful classrooms.
Photo credit: Hug It Forward

Kids in San Martín Jilotepeque, Guatemala, help collect and stuff plastic bottles to make ecobricks, which are used to build the walls of these schools.
Photo credit: Hug It Forward

"The Munching, Crunching Caterpillars" is a poem in *free verse*, which has no set meter or rhyme scheme.

"Be Straw Free" is a *triolet*, an eight-line poem in which line one repeats as lines four and seven and line two repeats as line eight. The rhyme scheme is ABaAabAB; the capital letters show lines that repeat.

Milo Cress delivered a congressional briefing about his "Be Straw Free" campaign on Capitol Hill, in Washington, DC.
Photo credit: Odale Cress

"A Shining Light" is a *cinquain*, a form of poetry composed of five lines with a pattern of two, four, six, eight, and two syllables.

"Stand Up, Speak Up" is a *concrete* or *shape poem*, as is "The Great Pacific Garbage Patch."

"Join the Crew" is a poem in *free verse*, as is "The Munching, Crunching Caterpillars."

The *Flipflopi*, a traditional Swahili dhow constructed from tons of plastic waste including flip-flops, sets sail from the Kenyan coast to raise awareness about plastic pollution.
Photo credit: Andras Porffy

For Further Reading

Children's Books
• *One Plastic Bag: Isatou Ceesay and the Recycling Women of the Gambia* by Miranda Paul; illustrations by Elizabeth Zunon. Minneapolis, MN: Millbrook Press, 2015.
• *Plastic Ahoy! Investigating the Great Pacific Garbage Patch* by Patricia Newman; photographs by Annie Crawley. Minneapolis, MN: Millbrook Press, 2014.
• *Tracking Trash: Flotsam, Jetsam and the Science of Ocean Motion* by Loree Griffin Burns. Boston: Houghton Mifflin, 2007.

News You Can Use

Annual Ocean Conservancy's International Coastal Cleanup
Each year, Ocean Conservancy conducts a coastal cleanup in more than 100 countries to provide current data on the state of our oceans. Here's the latest: www.oceanconservancy.org.

"Ten Tips to Reduce Your Plastic Footprint"
What you can do now according to the World Wildlife Fund: www.wwf.org.uk/updates/ten-tips-reduce -your-plastic-footprint

Caution about Wood
When using wood products, look for the label from the Forest Stewardship Council (FSC) that certifies the item is eco-friendly. Check www.us.fsc.org/en -US-/market/find-products.

Educator Guides
Kids Against Plastic: "Lesson Guides & Resources": How teachers can make students "plastic clever": www.kidsagainstplastic.co.uk/learn/lesson-guides

National Oceanic and Atmospheric Administration (NOAA) "Ocean Pollution": links to lesson plans, info, and videos: www.noaa.gov /education/resource-collections/ocean-coasts -education-resources/ocean-pollution

Recycling
Recycling regulations are determined by states and local municipalities, and new technology using optics and magnets is constantly changing how we recycle. So be sure to check your local recyclers for the latest dos and don'ts about what to place in your recycling bin.

Useful Websites
What do the numbers inside the triangular recycling symbol mean? Each identifies the type of plastic used, and not all can be recycled. Check with local recyclers to see which they can accept. For more info, see "Plastics by the Numbers": https://learn .eartheasy.com/articles/plastics-by-the-numbers

Need to know where to recycle items near you? Check: https://search.earth911.com

Wondering if something is biodegradable? Check the Biodegradable Products Institute: www.bpiworld.org

Acknowledgments

Special thanks to George H. Leonard, Ph.D., chief scientist, Ocean Conservancy, for fact-checking this book and for his expert advice. Join him and other scientists fighting for trash-free seas by visiting www.oceanconservancy.org.

Thanks to math whiz Jim Harman for checking my calculations converting metric measurements; to my friends and fellow poets in my writing group; to illustrator Christiane Engel; to my amazing agent Brenda Bowen; and to my dream team at HarperCollins—Nancy Inteli, Megan Ilnitzki, Chelsea C. Donaldson, Honee Jang, Caitlin Stamper, Nicole Moulaison, and Renée Cafiero. —S.H.